T0235859

SpringerBriefs in Electrical and Computer Engineering

Series Editors

Woon-Seng Gan, School of Electrical and Electronic Engineering, Nanyang Technological University, Singapore, Singapore

C.-C. Jay Kuo, University of Southern California, Los Angeles, CA, USA

Thomas Fang Zheng, Research Institute of Information Technology, Tsinghua University, Beijing, China

Mauro Barni, Department of Information Engineering and Mathematics, University of Siena, Siena, Italy

SpringerBriefs present concise summaries of cutting-edge research and practical applications across a wide spectrum of fields. Featuring compact volumes of 50 to 125 pages, the series covers a range of content from professional to academic. Typical topics might include: timely report of state-of-the art analytical techniques, a bridge between new research results, as published in journal articles, and a contextual literature review, a snapshot of a hot or emerging topic, an in-depth case study or clinical example and a presentation of core concepts that students must understand in order to make independent contributions.

More information about this series at http://www.springer.com/series/10059

Ricardo Moreno Chuquen · Harold R. Chamorro

Graph Theory Applications to Deregulated Power Systems

 Springer

Ricardo Moreno Chuquen
Universidad Autónoma de Occidente
Cali, Colombia

Harold R. Chamorro
KU Leuven
Leuven, Belgium

ISSN 2191-8112 ISSN 2191-8120 (electronic)
SpringerBriefs in Electrical and Computer Engineering
ISBN 978-3-030-57588-5 ISBN 978-3-030-57589-2 (eBook)
https://doi.org/10.1007/978-3-030-57589-2

This Springer imprint is published by the registered company Springer Nature Switzerland AG
The registered company address is: Gewerbestrasse 11, 6330 Cham, Switzerland

Contents

List of Figures

List of Tables

Chapter 1
Introduction

Abstract In this chapter, we describe the setting for the problems of interest for the work presented in this book. We first introduce the background and the motivation of our research. Then, we present the scope of this report and highlight the major contributions if this work. We end with an outline the contents of the remainder of this chapter.

1.1 Background and Motivation

Networks are present in almost every aspect of our life. The world surrounding us is full of networks. Social networks consist of relationships among social entities (i.e., friends, colleagues, relatives, etc.). Relationships among firms, vendors, manufacturers, and producers of raw material represent economic networks. On the other hand, technological networks such as the Internet, the computer data network that represents the interaction among computers connected by physical data connections. The World Wide Web network, associated with the Internet, consists of web pages connected by "hyperlinks". Also, Communication networks consist of telephones and mobiles phones. The focus of this book is on the power system networks that have evolved into the largest and most complex system of the technological age. Additionally, transportation system constitutes important networks i.e., cities and countries are connected by road or airline networks. Most recently, networks have become important in biology, in concrete to study three fields; the first one is related with biochemical networks, such as metabolic networks, protein-protein interaction networks, and genetic regulatory networks. A metabolic network is a representation of the chemical reactions that fuel cells and organism. The second one, it's the application of networks to gain insight into neural networks that consist of the interaction among neurons, ecological networks are the third field of study in which are studied the interaction among species i.e., predator-prey relationships.

The study of networks had a long history in mathematics and the sciences. Specifically, graph theory has been the mathematical language for describing the properties of networks. Graph theory is rooted in the eighteen century when the mathematician

R. Moreno Chuquen et al., *Graph Theory Applications to Deregulated Power Systems*,
SpringerBriefs in Electrical and Computer Engineering,
https://doi.org/10.1007/978-3-030-57589-2_1

Leonard Euler became interested in the Konigsberg Bridge Problem. Konigsberg was a flowering city in eastern Prussia which was built on the banks of the Pregel River, and on two islands that lie in midstream. Seven bridges connected the land masses. The people from Konigsberg amused themselves with mind puzzles, one of which was: "Can one walk across the seven bridges and never cross the same one twice?" Euler proved that this path does not exist. Euler's great insight lay in viewing the Konigsberg Bridge Problem as a graph by replacing each of the four land masses with nodes and each bridge with a link, obtaining a graph with four nodes and seven links. The proof is based on the number of links associated with each node; nodes with an odd number of links must be either the starting or the end point of the journey. Thus, a path crossing the seven bridges cannot exist because the Konigsberg graph had more than two nodes with an odd number of links [1] (Figs. 1.1 and 1.2).

Fig. 1.1 Graph example

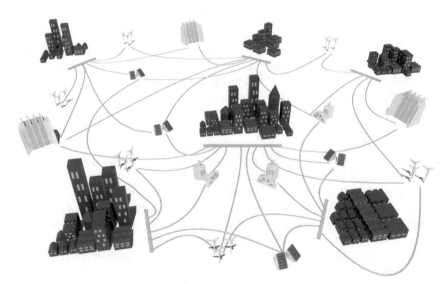

Fig. 1.2 Graph example

1.2 Scope and Contribution of This Book

The backbone of the power system grid is the transmission system whose main function is maintaining the connection among substations to supply electric energy with a certain level of security. The interconnection of several substations provides enough connecting paths to maintain the integrity of the grid. Some particular set of transmission lines allow sharing resources, such as generation reserves, which increases the reliability of the power grid. Therefore, the transmission system plays a major role in the continuous operation of the power grid. The security and reliability of the power systems depends to a large extent on the interconnectivity level of the power systems networks. However, the power grids are often not meshed enough because the investment in the construction of the transmission lines is very expensive. Thus, the quantification of structural properties can provide meaningful information needed to assess and enhance the reliability and security of power system networks.

We propose the extensively usage of graph theory to characterize the power system network structure. Graph theory offers a complete mathematical formulation which is very useful to capture the structure of any graph. We develop a comprehensive characterization of the networks associated to power systems based on the graph theory mathematics but considering the intrinsic and unique structural properties of the power systems.

Specifically, we deduce valuable results from spectral graph theory to study and gain insight in the topological properties of power networks. Spectral graph theory is a study of the spectra of matrices that characterize properties of a graph. Matrices like the degree matrix, adjacency matrix, and the Laplacian matrix have eigenvalues and eigenvectors that can give important information about the structure of the system and its intrinsic robustness.

1.3 Book Outline

This book contains six additional chapters. We start with a complete topological characterization of the structure of the power system networks in Chap. 2. We use extensively graph theoretic concepts to capture the structure of the power system networks. We propose a modified version of the admittance matrix which is similar, mathematically speaking, to the structure and properties of the laplacian matrix.

In Chap. 3, we formulate the identification of k subnetworks in a power system network as an unconstrained optimization problem. We use the mathematical representation deduced in the Chap. 2 to formulate the problem, we find out that the solution of this problem is on the eigenspace of the Laplacian matrix that represents the power grid.

We deduce an index to quantify the network robustness in Chap. 4. We provide a concept of network robustness based on the topological firmness of the interconnection among substations through of the transmission system. We study the network robustness of the IEEE 118-bus system based on the index deduced and we find out interesting results about the structure of this particular system.

In Chap. 5, we develop a tool which is focused on security strategies. We call this strategy as "Hierarchical Islanding". This tool was developed as a software tool based mainly on two modules: Topological Module (TM) and Spectral Analysis Module (SAM). The strategy offers the possibility to power system operators to eject an islanding-action to mitigate grid-impactive events. Additionally, this chapter presents some security studies based on the critical tie lines connecting subnetwork using the so-called Power Transfer Distribution Factors (PTDFs).

In Chap. 6, we discuss about cyber physical system security. This topic has received intense attention from the power system society during the last years. We discuss about different types of attacks to Energy Management System (EMS) and about the impact of these attacks over the operation of the power systems including the electricity markets.

Concluding remarks are provided in Chap. 7. We summarize the work presented together with discussion directions for future research to extend the results in this work.

This book has three appendices. Appendix A.1 provides a summary of the acronyms and the notation used in the book. In Appendix , we provide both the statement of an equation and the solution of a quadratic optimization problem used in the Chap. 3.

Reference

1. Barabási A-L, Linked: The New Science of Networks. Perseus Publishing, Cambridge, 2002

Chapter 2
Topological Characterization of Power Systems Networks

Abstract This chapter presents detailed graph theoretic concepts with emphasis on applications on power system problems. Since the interest is in the disclosure of important network properties, it considers the construction of the adjacency, degree and the Laplacian matrices and the properties that can be deduced from them. Specifically, this chapter deduces results from the eigenvalues and eigenvectors of the adjacency and Laplacian matrices related with network properties. Moreover, this chapter includes a complete explanation and deduction from an optimization point of view for all results and his insights for applications.

We consider a power system consisting of $N + 1$ buses and E lines. We denote by $N = \{0, 1, \ldots, N\}$ the set of network nodes, with the bus 0 being the slack bus and by $\mathcal{L} = \{l_1, l_2, \ldots, l_E\}$ the set of transmission lines lines and transformers that connect the nodes in the set N. Each line $l_m \in N, m = \{1, \ldots, E\}$ is associated with an unordered pair of nodes (i, j). For power system networks we do not consider any self-loops associated with a node pair (i, i). The network may have two or more transmission lines associated with the same node pair (i, j) i.e., parallel transmission lines. We associate the undirected graph $\mathcal{G} = \{N, \mathcal{L}\}$ with the power system network. The graph $\mathcal{G}_w = \{N_w, \mathcal{L}_w\}$ is a subgraph of $\mathcal{G} = \{N, \mathcal{L}\}$ whenever all nodes $N_w \subset N$ and all the transmission lines $\mathcal{L}_w \subset \mathcal{L}$. We use the terms graph and network interchangeably in the remainder of this chapter.

We characterize algebraically the graph $\mathcal{G} = \{N, \mathcal{L}\}$ by the connectivity, the degree, and the Laplacian matrices, denoted by A, D and, L respectively. A is a $(N + 1) \times (N + 1)$ matrix whose elements indicate the number of lines that directly connected the node pair (i, j) [1]. Whenever the node pair (i, j) is not directly connected the corresponding element of A, $a_{i,j}$ is equal to zero. By definition $\mathbf{A} = A^T$. The degree d_i of the node $i \in N, i = \{0, 1, \ldots, N\}$, is defined as the number of transmission lines incident with the bus i.

We construct $\mathbf{D}$ to be the diagonal matrix,

$$\mathbf{D} = \text{diag}\,\{d_0, d_1, \ldots, d_n\} \tag{2.1}$$

© The Author(s), under exclusive license to Springer Nature Switzerland AG 2021
R. Moreno Chuquen et al., *Graph Theory Applications to Deregulated Power Systems*,
SpringerBriefs in Electrical and Computer Engineering,
https://doi.org/10.1007/978-3-030-57589-2_2

The Laplacian matrix $\mathbf{L}$ provides a mathematical relationship between the matrices $\mathbf{A}$ and $\mathbf{D}$ since,

$$\mathbf{L} = \mathbf{D} - \mathbf{A} \tag{2.2}$$

Clearly, the elements of $\mathbf{L}$ are

$$\mathbf{L}_{i,j} = \begin{cases} d_i = v_i^d i_i^d \\ -a_{ij} = v_i^d i_i^d \\ 0 = -v_i^d i_i^d \end{cases} \tag{2.3}$$

$\mathbf{L}$ provides a succinct measure of the connectivity in $\mathcal{G}$. We make extension use of the useful algebraic properties of $\mathbf{L}$ [2]. The matrix $\mathbf{L}$ is by inspection, symmetric and $\mathbf{L}$ is singular because $\mathbf{L1} = 0$ where $\mathbf{1} = [1, 1, \ldots, 1]^T$. This property indicates that $\mathbf{L}$ has at least one zero eigenvalue. Moreover, $\mathbf{L}$ is positive semidefinite, all its eigenvalues are non-negative. We order the $(N + 1)$ eigenvalues λ_i of $\mathbf{L}$ in nondecreasing order $\lambda_1 \leq \lambda_2 \leq \lambda_3 \leq \cdots \leq \lambda_{N+1}$. Additional properties of $\mathbf{L}$ are given in [3, 4].

We consider the graph $\mathcal{G} = \{\mathcal{N}, \mathcal{L}\}$ graph that represents a power network and its constitute subgraphs $\mathcal{G}_1 = \{\mathcal{N}_1, \mathcal{L}_1\}, \mathcal{G}_2 = \{\mathcal{N}_2, \mathcal{L}_2\}, \mathcal{G}_{\mathcal{G}} = \{\mathcal{N}_{\mathcal{G}}, \mathcal{L}_{\mathcal{G}}\}$. The set of $\mathcal{G}$ subnetworks $\mathcal{G}_1 = \{\mathcal{N}_1, \mathcal{L}_1\}, \mathcal{G}_2 = \{\mathcal{N}_2, \mathcal{L}_2\}, \ldots, \mathcal{G}_k = \{\mathcal{N}_k, \mathcal{L}_k\}$ of $\mathcal{G}$ are disjoint if $\mathcal{N}_1 \cap \mathcal{N}_2 \cap \cdots \cap \mathcal{N}_k = \emptyset$ with and $\mathcal{N}_1 \cup \mathcal{N}_2 \cup \cdots \cup \mathcal{N}_k = \mathcal{N}$. The elements of $\mathcal{L} \cap$ are the interconnecting lines of the $\mathcal{G}$ subnetworks. Each such line is characterized in terms of its node pair (i, j) with $i \in \mathcal{N}_k$ and $j \in \mathcal{N}_{k'}, k \neq k'$.

The connectivity of the graph $\mathcal{G} = \{\mathcal{N}, \mathcal{L}\}$ and its constituent subnetworks $\mathcal{G}_1 = \{\mathcal{N}_1, \mathcal{L}_1\}, \mathcal{G}_2 = \{\mathcal{N}_2, \mathcal{L}_2\}, \ldots, \mathcal{G}_k = \{\mathcal{N}_k, \mathcal{L}_k\}$ may be deduced from its spectral information. The eigenvalues and eigenvectors of the Laplacian matrix $\mathbf{L}$ reveal intrinsic information about structural properties of the power networks which could be used extensively in a wide range of security problems. The connectivity of a graph $\mathcal{G}$ is formally defined in terms of paths that connect all the buses $\mathcal{N}$.

Definition 2.1 A graph $\mathcal{G}$ is connected if and only if for every node pair there exist a path from i to j [5].

Matrix $\mathbf{L}$ has been defined as a real, symmetric and positive semidefinite that measures the connectivity among buses. During the development of this book we propose selecting a modified version of the impedance matrix as Laplacian matrix in virtue that both matrices have the same mathematical structure. The laplacian matrix $\mathbf{L}$ could be taken as a real approximation of the impedance matrix $\mathbf{Z}_{bus}$. The inverse of $\mathbf{Z}_{bus}$ corresponds to the admittance matrix $\mathbf{Y}_{bus} = \mathbf{G}_{bus} + \mathbf{B}_{bus}$ which is sparse. For practical transmission power systems the admittance matrix can be approximated by the susceptance matrix $\mathbf{Y}_{bus} \approx \mathbf{B}_{bus}$. The matrix $\mathbf{B}_{bus}$ is real, symmetric and negative semidefinite. The matrix $\mathcal{L}$ may be taken as the negative of the inverse susceptance matrix $\mathbf{B}_{bus}^{-1}$ which is real, symmetric and positive semidefinite.

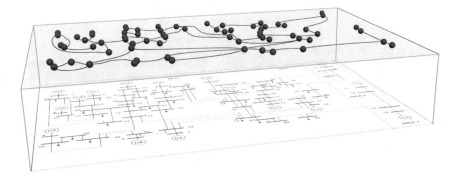

Fig. 2.1 Topological characterization of the 68-bus system

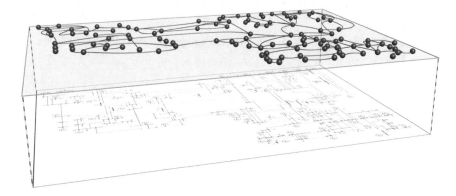

Fig. 2.2 Topological characterization of the IEEE 118-bus test system

This book will illustrate various concepts using standard electrical systems. We will choose a medium scale system as the 68-bus system and a large scale system as IEEE 118-bus system to study different applications. The 68-bus test power system is a reduced order equivalent of the interconnected New England test system (NETS) and New York Power System (NYPS). There are five geographical regions; the areas 3, 4 and 5 represent equivalent dynamical systems [6]. The corresponding topological structure of this system represented as a graph is showed in the Fig. 2.1. The IEEE 118-bus system model described in [7] is a standard electrical system well known in the research community; and its topological characterization is illustrated in the Fig. 2.2

References

1. Linked: the new science of networks, 1st edn. Perseus Books Group
2. Godsil C, Royle GF, Algebraic graph theory. Graduate texts in mathematics, Springer, Berlin. https://www.springer.com/gp/book/9780387952413. https://doi.org/10.1007/978-1-4613-0163-9
3. Chung F, Laplacians and the Cheeger inequality for directed graphs 9(1):1–19. http://dx.doi.org/10.1007/s00026-005-0237-z
4. Spielman DA (2010) Algorithms, graph theory, and linear equations in laplacian matrices. In: Proceedings of the international congress of mathematicians 2010 (ICM 2010), Published by Hindustan Book Agency (HBA), India. WSPC Distribute for All Markets Except in India, pp 2698–2722
5. Bollobas B, Modern graph theory. Graduate texts in mathematics, Springer, Berlin. https://www.springer.com/gp/book/9780387984889, https://doi.org/10.1007/978-1-4612-0619-4
6. Pal B, Chaudhuri B, Robust control in power systems power electronics and power systems. Springer, US. https://www.springer.com/gp/book/9780387259499. https://doi.org/10.1007/b136490
7. [Online] Available: http://www.ee.washington.edu/research/pstca/pf118/pg_tca118fig.htm

Chapter 3
Identification of Multiple Subnetworks

Abstract The structure of the power system networks is a function of the status of each transmission line. Our focus in this chapter is on the topological assessment with the objective to develop a practical scheme to identify the tie lines that connect the constituent sub-networks in the grid structure. We develop a useful topological characterization of the power system network based on graph theoretic concepts to gain insights into its structural properties. The identification of multiple sub-networks provides effective schemes to analyze critical contingencies.

Network structural analysis has been applied to study a wide range of applications in different fields. Research in biology used a graph theoretic approach to characterize the structure of molecules [1, 2]. The study of the World Wide Web network as a graph allows the identification of useful connectivity properties [3, 4]. The structure of the mobile communication network is analyzed to evaluate its connectivity [5]. The broad range of interactions between individuals in society is well modeled as a social network [6]. The focus of this chapter is on the structural analysis of the power system network topology.

There exist various approaches to identify subnetworks and the tie lines that interconnect them. The identification of tie lines offers practical information for the security assessment of the power system [7, 8]. Some papers have proposed search techniques to identify cut-sets based on a slow coherency approach [9–11]. Simulated annealing is used to identify subnetworks in order to reduce the use of parallel computers systems [12]. Another approach uses a reduce network of the original power system network searching for subnetworks based on decision diagrams [13]. The identification of observable islands in an unobservable network is studied in [14]; the boundaries are defined based on the monitoring of line flows. The transmission lines near to its full capacity and geographically aspects are taken into consideration to identify tie lines in the proposal of [15]. Although these schemes to identify subnetworks or islands use graph-theoretic concepts, they use simplified graph models.

R. Moreno Chuquen et al., *Graph Theory Applications to Deregulated Power Systems*,
SpringerBriefs in Electrical and Computer Engineering,
https://doi.org/10.1007/978-3-030-57589-2_3

These kinds of simplifications involve missing of information about the structural properties of power system networks. Combinatorial methods faced a computationally expensive issue because are based on the evaluation of several scenarios. In this chapter, we address the need of a complete topological characterization of power system networks and so the identification of multiple subnetworks based on spectral information about the grid. The scheme proposed in this paper is useful to perform effectively security analysis because the identification of tie lines is a critical property studied for power system security [16]. Each power system network may be conceived as the arrangement of various subnetworks which are connected between them through of specific tie lines. These tie lines play an important role in the network structure since the removal leads to island formation [17]. In other words, the lines connecting a particular subnetwork to the whole network reflect the robustness of the system with respect to disturbances [18].

We propose the development of a graph theoretic approach to assess the power system network from a structural perspective. A range of problems in the area of power system security may be addressed using information deduced from the topological analysis of the power system network. Specifically, the identification of transmission lines connecting subnetworks gives us useful information about the structural function of tie lines. Each network has some specific set of transmission lines whose structural function is to maintain the connection of the whole network. Thus, the removing of one of these lines is more critical for the system in terms of the connectivity than the removing of any other line. For security studies, the impacts of tie lines outages into the power system network could be quantified using the information deduced from the topological assessment. We use the so-called power transfer distribution factors (PTDFs) to quantify the impact of line outages on the non-outaged lines' flows.

3.1 Identification of Two Subnetworks

To start out, we study the problem of the identification of two subnetworks of a power system. We define $\mathbf{r}$ as the indicator vector, whose elements r_i we associate with each node $i \in \mathcal{N}$ of the power system, where $r_i = +1$ if bus i belongs to subnetwork 1 or $r_i = -1$ if the bus i belongs to subnetwork 2. Note that $\mathbf{r}$ satisfies $\mathbf{r}^T \mathbf{r}$. If we assume a particular indicator vector $\mathbf{r}$ to identify two subnetworks of $\mathcal{G}$, represented by two graphs $\mathcal{G}_1 =$ and $\mathcal{G}_2 =$, then there is a particular set of transmission lines $\mathcal{L}_{12} =$ connecting the subnetworks $\mathcal{G}_1$ and $\mathcal{G}_2 =$ which can be expressed in terms of the external connections $\mathcal{L}_1$ and as The cardinality of can be quantified using the indicator vector

$$\mathbf{L}_{i,j} = \frac{1}{2} \sum_{i,j \in \mathcal{N}} \left(r_i - r_j \right)^2 \tag{3.1}$$

The quadratic form of the matrix $\mathbf{L}$ is $\mathbf{r}^T\mathbf{L}\mathbf{r}$ which is constructed using particularly the indicator vector $\mathbf{r}$. We show in Appendix that the expression in (3.1) may be written using the quadratic form of $\mathbf{L}$, given as follows

$$\frac{1}{2}\sum_{i,j\in N}\left(r_i - r_j\right)^2 = \mathbf{r}^T\mathbf{L}\mathbf{r} \tag{3.2}$$

The only non-zero terms in the sum in (3.2) are those that correspond to the set of transmission lines that connect a node in network $\mathcal{G}_1$ to a node in network $\mathcal{G}_2$. The identification problem determines the set of transmission lines $\mathcal{L}_{12}$ connecting $\mathcal{G}_1$ and $\mathcal{G}_2$, whose union constitutes $\mathcal{G}$, by solving the quadratic problem as follows:

$$min\left(\mathbf{r}^T\mathbf{L}\mathbf{r}\right) \tag{3.3}$$

subject to,

$$\mathbf{r}^T\mathbf{r} = N \tag{3.4}$$

$$\mathbf{1}^T\mathbf{r} = q \tag{3.5}$$

$$r_i = \pm 1 \tag{3.6}$$

The condition in (3.4) affirms that every node is assigned to one of the two subnetworks, and the condition (3.5) specifies that the number of nodes inside one subnetwork is equal to q. The solution of the optimization is presented in Appendix and is given by the vector $\mathbf{r}^*$ that solves,

$$(\mathbf{L} - \lambda\mathbf{I})\,\mathbf{r}^* = 0 \tag{3.7}$$

For the second smallest eigenvalue λ of $\mathcal{L}$, $\mathbf{r}^*$ is the corresponding eigenvector. We consider the solution $\mathbf{r} =$ to be trivial and it corresponds to the subnetwork $\mathcal{G}_1$ being the entire network and the empty subnetwork $\mathcal{G}_2 = \emptyset\emptyset$, while such a partition of the network is valid it is not useful. This trivial solution corresponds to the smallest eigenvalue $\lambda = 0$.

For $\mathbf{r}^T\mathbf{L}\mathbf{r}$ to be minimum, we select the eigenvalue λ_2, i.e., the smallest nonzero magnitude eigenvalue. Consequently, $\mathbf{r}^*$ is an eigenvector corresponding to λ_2. The second smallest eigenvalue λ_2 provides information on the network partitioning into the subnetworks $\mathcal{G}_1$, and its complement $\mathcal{G}_2$. Therefore, the indicator vector $\mathbf{r}$ must be chosen to be a scaled value of the eigenvector $\mathbf{r}^*$. Choosing $\mathbf{r}$ to be as close to parallel with $\mathbf{v}_2$ as possible provides the solution [19]. This means maximizing the dot product:

$$\left|\mathbf{v}_2^T\mathbf{r}\right| = \left|\sum_i \mathbf{v}_2 r_i\right| = \sum_i \mathbf{v}_2 \tag{3.8}$$

The second relation follows by the triangle inequality and becomes equality only when all terms in the first sum are either positive or negative. The maximum of $\mathbf{v}_2^T \mathbf{r}$ is reached when $\mathbf{v}_{2i}\mathbf{r}_i \geq 0$ for all i, or equivalently, when r_i has the same sign as the component i of $\mathbf{v}_2$. Then, the maximum is obtained under the condition [19]:

$$r_i = \begin{cases} +1 = \mathbf{v}_2 \\ -1 = \mathbf{v}_2 \end{cases} \tag{3.9}$$

This development has shown that the identification problem of a graph is transformed into an eigenvalue and eigenvector problem involving the graph Laplacian. The spectral information extracted from the Laplacian matrix allows the direct identification of two islands.

3.2 Algebraic Connectivity

There is an alternative approach to demonstrate the interesting properties given by the spectral information deduced of the Laplacian matrix. The properties of the second eigenvalue λ_2 and the corresponding eigenvector $\mathbf{v}_2$ of the Laplacian matrix reveal connectivity properties of a graph how was deduced before. The eigenpair $(\lambda_2, \mathbf{v}_2)$ was called the Algebraic Connectivity of $\mathcal{G}$ by Fiedler [20, 21]. The following is a summary of his observations. If $\mathbf{L}$ is positive semidefinite matrix then the second smallest eigenvalue is equal to:

$$\lambda_2 = min \left(\frac{\mathbf{v}_2 \mathbf{L} \mathbf{v}_2}{\mathbf{v}_2^T \mathbf{v}_2} \right) \tag{3.10}$$

by the Courant-Fischer theorem [22]. Let $\mathcal{G}$ be a connected graph, and let $\mathbf{v}_2$ be the eigenvector corresponding to λ_2. The subgraphs induced by each one of the conditions of (12) are connected [21].

The components of the second eigenvector are assigned to the vertices of $\mathcal{G}$; so that each node of an interconnected power system will be assigned a value based on the corresponding value of the second eigenvector of the Laplacian matrix (that is, if the node has number k, the k_{th} component of the second eigenvector will be associated with it).

The Cheeger's inequality [23] provides an explanation why the second eigenvalue and the corresponding eigenvector can be used for efficiently partitioning a graph. The graph-theoretic version of the Cheeger's inequality presented in this paper is based on [24, 25]. The graph bisection problem is to find a set of buses $\mathcal{N}_w$ such that is a subset of $\mathcal{N}$, in notation such that the number of buses in $\mathcal{N}_w$ is almost half of $\mathcal{N}$

$$\mathcal{N}_w \subseteq \frac{|\mathcal{N}|}{2} \tag{3.11}$$

(the operator $|.|$ refers to cardinality), in fact minimizing the set of cut edges:

$$(i, j) \in \tag{3.12}$$

To find a minimal cut, we will introduce a cut ratio,

$$\varphi(\mathcal{G}) \equiv \frac{|\partial \mathcal{G}_w|}{\partial \mathcal{N}_w \partial (\mathcal{N} - \mathcal{N}_w)} \tag{3.13}$$

where $\partial(\mathcal{G}_w)$ is called the boundary of $\mathcal{N}_w$ it denotes the number of edges with one end point in $\mathcal{G}_w$ and the other in $\mathcal{G} - \mathcal{G}_w$ is defined by:

$$d(\mathcal{N}_w) \equiv \sum_{i \in \mathcal{N}_w} d(i) \tag{3.14}$$

where $d(i)$ is the degree of the vertex i. The cut ratio is also called the conductance. The partitioning problem is to find the set of lines with minimum conductance. This minimum defines the conductance of a graph $\mathcal{G}$ thus:

$$\Phi(\mathcal{G}) = min\varphi(\mathcal{N}_w) \tag{3.15}$$

The graph conductance $\Phi(\mathcal{G})$ has bounds given by the Cheeger's inequality:

$$\frac{\lambda_2}{d(\mathcal{N})} \leq \Phi(\mathcal{G}) \leq \frac{2\sqrt{2\lambda_2}}{d(\mathcal{N})} \tag{3.16}$$

The graph theoretic version of (11) is [25]:

$$\Phi(\mathcal{G}) \leq 2\sqrt{2\lambda_2} \tag{3.17}$$

3.3 Identification of K-Subnetworks

The extension of the formulation to identify multiple islands into power systems can be modeled as an optimization problem with a quadratic objective function extended to k dimensions. The objective function may be written as:

$$min \left\{ \sum_{i=1}^{k} \mathbf{r}_i^T \mathbf{L} \mathbf{r}_i \right\} \tag{3.18}$$

where each $\mathbf{r}_i$ indicator vector assigns a coordinate for each bus into the k-dimensional space. This problem is constrained by the orthogonality condition:

$$\mathbf{r}^T \mathbf{r}_i = 1 \tag{3.19}$$

The mathematical solution is obtained following the procedure explained in Appendix . The solution is given by:

$$\mathbf{L} - \lambda_i \mathbf{I} = 0 \tag{3.20}$$

The solution is a nontrivial if λ_i is nonzero eigenvalue of $\mathbf{L}$ and $\mathbf{r}_i$ is its corresponding eigenvector. The premultiplication of (3.13) by $\mathbf{r}_i^T$ with the constraint of (15) holding results in:

$$\lambda_i = \mathbf{I} = \mathbf{S}^T \mathbf{S} \tag{3.21}$$

If follows that the minimum value of (3.11) results for the sum of the k smallest nonzero eigenvalues. The corresponding k eigenvectors are the solution to identify k islands. We construct the matrix $\mathbf{S} \in$, whose columns are the k normalized eigenvectors of the matrix $\mathbf{L}$ corresponding with the k smallest nonzero eigenvalues. The matrix $\mathbf{S}$ is an orthogonal matrix, satisfying:

$$\mathbf{S}\mathbf{S}^T = \mathbf{I} = \mathbf{S}^T \mathbf{S} \tag{3.22}$$

Where, $\mathbf{I}$ is the identity matrix. If, the matrix Λ is the diagonal matrix of nonzero eigenvalues,

$$\Lambda = diag\,[\lambda_2, \lambda_3, \ldots, \lambda_N] \tag{3.23}$$

Then, the singular value decomposition for $\mathbf{L}$ is given by the orthogonal matrix $\mathbf{S}$ and the diagonal matrix Λ,

$$\mathbf{S}\mathbf{L}\mathbf{S}^T = \Lambda \tag{3.24}$$

The node i is associated with the row i of the matrix $\mathbf{S}$ to assign each node to a particular subnetwork k. The row i is interpreted as the coordinates of the bus i in, each node of the power system network is associated a vector represented by the row i. We use the squared Euclidean distance to decide the membership of each bus to a particular subnetwork k. The usage of this criterion based on the spectral information obtained from the Laplacian matrix gives us insight about the relative position of each node inside of the power system network. The vector i associated to the bus i with a high relative value of squared Euclidean distance implies that the corresponding bus is far away to the boundaries with other subnetworks. Meanwhile, small relative value of the squared Euclidean distance implies that the corresponding bus is close to the boundaries with other subnetworks.

3.4 Illustrative Examples: the IEEE 118-Bus System

We illustrate the application of the identification of multiple islands. Security analyses are developed to evaluate the static security of the power systems based on the identification of k-islands. In the connected IEEE 118-bus test system, we identify ten islands considering the status of each line to determine the system connectivity,

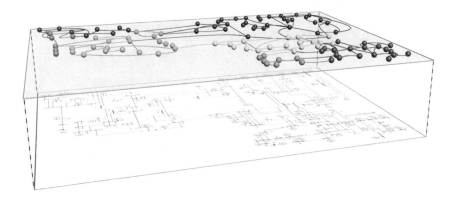

Fig. 3.1 Ten Islands identified for the graph model of IEEE 118-bus test system

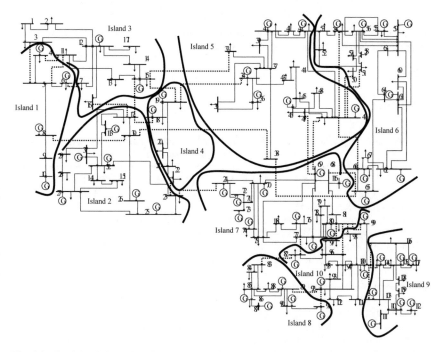

Fig. 3.2 Ten Islands identified for the IEEE 118-bus test system

Table 3.1 Tie transmission lines for the IEEE 118-bus system

Set TL$_1$	{(3, 5), (4, 11), (5, 11), (6, 7), (8, 30)}
Set TL$_2$	{(17, 16), (17, 15), (30, 38), (23, 22), (23, 24), (30, 8)}
Set TL$_3$	{(3, 5), (4, 11), (5, 11), (6, 7), (16, 17), (15, 17), (15, 19), (15, 33)}
Set TL$_4$	{(19, 15), (19, 34), (18, 17), (22, 23)}
Set TL$_5$	{(33, 15), (34, 19), (38, 30), (38, 65), (49, 66), (49, 51), (49, 50), (49, 54), (49, 69), (47, 69)}
Set TL$_6$	{(54, 49), (50, 49), (51, 49), (66, 49), (65, 116), (65, 38)}
Set TL$_7$	{(68, 65), (69, 49), (69, 47), (24, 23), (77, 82), (80, 97), (80, 96), (80, 98), (80, 99)}
Set TL$_8$	{(83, 82), (89, 92), (91, 92)}
Set TL$_9$	{(106, 100), (104, 100), (103, 100)}
Set TL$_{10}$	{(99, 80), (98, 80), (96, 80), (92, 89), (97, 80), (82, 77), (82, 83), (92, 91), (100, 106), (100, 104), (100, 103)}

the construction of the matrix is based whether or not a line in the power system network is connected. We construct the matrix **L** based on the eigenvectors of **L** corresponding to the first ten smallest nonzero eigenvalues of **L**. Figure 3.1 show ten islands identified for the graph model of IEEE 118-bus system. Figure 3.2 show the ten subnetwork in the IEEE 118-bus electrical system. The tie transmission lines that allow the identification of ten islands are summarized in the Table 3.1. The sets of transmission lines denoted by TL$_1$, TL$_2$, TL$_3$, TL$_4$, TL$_5$, TL$_6$, TL$_7$, TL$_8$, TL$_9$ and TL$_{10}$ are the tie lines connecting each island with the whole network. These results allow the deduction of relevant information for security analysis, for instance, we observe that the islands 8 and 9 are connected individually with the whole network through of three transmission lines while the island 10 is connected with the whole network through of eleven transmission lines.

References

1. Gavin A-C, Aloy P, Grandi P, Krause R, Boesche M, Marzioch M, Rau C, Jensen LJ, Bastuck S, Dümpelfeld B, Edelmann A, Heurtier M-A, Hoffman V, Hoefert C, Klein K, Hudak M, Michon A-M, Schelder M, Schirle M, Remor M, Rudi T, Hooper S, Bauer A, Bouwmeester T, Casari G, Drewes G, Neubauer G, Rick JM, Kuster B, Bork P, Russell RB, Superti-Furga G, Proteome survey reveals modularity of the yeast cell machinery 440(7084):631–636. https://doi.org/10.1038/nature04532
2. Jeong H, Mason SP, Barabási A-L, Oltvai ZN, Lethality and centrality in protein networks 411(6833):41–42. http://dx.doi.org/10.1038/35075138
3. Broder A, Kumar R, Maghoul F, Raghavan P, Rajagopalan S, Stata R, Tomkins A, Wiener J, Graph structure in the web 33(1-6):309–320. http://dx.doi.org/10.1016/S1389-1286(00)00083-9
4. Faloutsos M, Faloutsos P, Faloutsos C, On power-law relationships of the Internet topology. In: SIGCOMM '99. http://dx.doi.org/10.1145/316188.316229
5. Onnela J-P, Saramäki J, Hyvönen J, Szabó G, Lazer D, Kaski K, Kertész J, Barabási A-L, Structure and tie strengths in mobile communication networks 104(18):7332–7336. http://arxiv.org/abs/17456605. http://dx.doi.org/10.1073/pnas.0610245104. https://www.pnas.org/content/104/18/7332

6. Social network analysis: methods and applications, 1st edn, Cambridge University Press, Cambridge
7. Ejebe GC, Wollenberg BF, Automatic contingency selection PAS-98(1):97–109. http://dx.doi. org/10.1109/TPAS.1979.319518
8. Stott B, Alsac O, Monticelli A, Security analysis and optimization 75(12):1623–1644. http:// dx.doi.org/10.1109/PROC.1987.13931
9. You H, Vittal V, Wang X, Slow coherency-based islanding 19(1):483–491. http://dx.doi.org/ 10.1109/TPWRS.2003.818729
10. Xu G, Vittal V, Slow coherency based cutset determination algorithm for large power systems 25(2):877–884. http://dx.doi.org/10.1109/TPWRS.2009.2032421
11. Chow JH, Slow coherency and aggregation. In: Chow JH (ed), Power system coherency and model reduction, power electronics and power systems. Springer, Berlin, pp 39–72
12. Irving M, Sterling M, Optimal network tearing using simulated annealing 137(1):69–72. http:// dx.doi.org/10.1049/ip-c.1990.0010
13. Sun K, Zheng D-Z, Lu Q, Splitting strategies for islanding operation of large-scale power systems using OBDD-based methods 18(2):912–923. http://dx.doi.org/10.1109/TPWRS.2003. 810995
14. Monticelli A, Wu FF, Network observability: identification of observable islands and measurement placement PAS-104(5):1035–1041. http://dx.doi.org/10.1109/TPAS.1985.323453
15. Dola H, Chowdhury B, Intentional islanding and adaptive load shedding to avoid cascading outages. In: 2006 IEEE power engineering society general meeting, p. 8. http://dx.doi.org/10. 1109/PES.2006.1709349
16. Meliopoulos A, Cheng C, Xia F, Performance evaluation of static security analysis methods 9(3):1441–1449. http://dx.doi.org/10.1109/59.336119
17. Guler T, Gross G, Detection of Island formation and identification of causal factors under multiple line outages 22(2):505–513. http://dx.doi.org/10.1109/TPWRS.2006.888985 https:// doi.org/10.1109/TPWRS.2006.888985
18. Moreno R, Torres A, Security of the power system based on the separation into islands. In: 2011 IEEE pes conference on innovative smart grid technologies latin America (ISGT LA), pp 1–5. http://dx.doi.org/10.1109/ISGT-LA.2011.6083210
19. Newman MEJ, Finding community structure in networks using the eigenvectors of matrices 74(3):036104. http://dx.doi.org/10.1103/PhysRevE.74.036104
20. Fiedler M, Algebraic connectivity of graphs 23(2):298–305. https://dml.cz/handle/10338. dmlcz/101168
21. Fiedler M, A property of eigenvectors of nonnegative symmetric matrices and its application to graph theory 25(4):619–633. https://dml.cz/handle/10338.dmlcz/101357
22. Courant R, Hilbert D, Methods of Mathematical Physics, Interscience, New York, Vol. 1, 1953
23. Cheeger J, A lower bound for the smallest eigenvalue of the laplacian, pp 195–199. https:// nyuscholars.nyu.edu/en/publications/a-lower-bound-for-the-smallest-eigenvalue-of-the-laplacian
24. Spectral graph theory (CBMS regional conference series in mathematics, No. 92) - PDF free download. https://epdf.pub/spectral-graph-theory-cbms-regional-conference-series-in-mathematics-no-92.html
25. Spielman DA, Spectral graph theory and its applications. In: 48th annual IEEE symposium on foundations of computer science (FOCS'07), pp 29–38. http://dx.doi.org/10.1109/FOCS. 2007.56

Chapter 4
Network Robustness for Power Systems

Abstract The identification of critical transmission lines and critical substations in power systems is prerequisite for security assessment studies because it provides effective schemes to analyze critical contingencies. Sometimes the power network exhibit high vulnerability related with critical transmission lines interconnecting critical substations from a physical point of view. This chapter describes a method to quantify the relative importance level of each substation for the connectivity of the whole network. Vulnerability studies are performed effectively using information about critical tie lines and about critical substations. Additionally, this chapter presents some security studies based on the critical tie lines connecting subnetwork using the so-called Power Transfer Distribution Factors (PTDFs).

While the topology information is, in principle, simple to understand, the determination of a robustness criterion is more complicated. This chapter provides a network index to quantify the topological robustness of power system networks based on the development of a graph theoretic approach to characterize the topological structure. The robustness criterion introduced in this chapter provides useful information to deal with a range of problems in the context of transmission system expansion.

The backbone of the power system grid is the transmission system whose main function is maintaining the connection among substations to supply electric energy with a certain level of security. The interconnection of several substations provides enough connecting paths to maintain the integrity of the grid. Some particular set of transmission lines allow sharing resources, such as generation reserves, which increases the reliability of the power grid. Therefore, the transmission system plays a major role in the continuous operation of the power grid [1]. The security and reliability of the power systems depends to a large extent on the inter-connectivity level of the power systems networks [2]. However, the power grids are often not meshed enough because the investment in the construction of the transmission lines is very expensive. Thus, the quantification of structural properties can provide meaningful

R. Moreno Chuquen et al., *Graph Theory Applications to Deregulated Power Systems*,
SpringerBriefs in Electrical and Computer Engineering,
https://doi.org/10.1007/978-3-030-57589-2_4

information needed to assess and enhance the reliability and security of power system networks.

The modeling and analysis of networks in terms of various topological metrics has attracted recently considerable attention [3, 4]. The vulnerability of the power system networks to the nodes and lines failures is an intrinsic structural property that reveals information about the topological firmness. Quantification of the topological robustness of a network allows us to understand how the power grid is constructed [5]. A suitable measure of robustness could be useful in making decisions related to the reinforcement of the structure. This paper proposes a robustness index based on the topological properties of the power system networks to quantify the firmness of the interconnection of the substations through the transmission system.

There exist methods in pure graph theory to quantify the robustness of a network. These approaches use the link and node connectivity metrics to obtain robustness measures [6]. The relationship between the minimum degree and the link and node connectivity is used to define an optimal graph [7, 8]; if the graph contains trees, then the metric does not provide accurate results. These methods require the computation of polynomial time algorithms to find the node and link connectivity [9]. Every change in the topology requires a new application of the algorithms starting from the "scratch". Therefore, the analysis of the topological robustness with graphs changing their structure requires repeated application of the topology based algorithms, which may result in computational inefficiencies. This paper proposes a direct quantification of network robustness in a computationally efficient way, developing a suitable topological characterization of the power system networks based on the graph theoretic concepts.

Topological information about power systems is usually available from the NTP (Network Topology Processor). The NTP module continuously retains and updates electrical system topological information such as node-branch network connectivity, open-ended lines and transformers. There exist various methods to perform this task. The pioneering work [10] proposes a method to determine network topology in real-time. A topology processor that tracks network modification in real-time is proposed in [11]. This chapter proposes the usage of the topological information to quantify the network robustness considering changes in the inter-connectivity.

This chapter has three more sections. Then, in Sect. 0, the robustness index is introduced. In Sect. 4.2, the proposed approach is used to investigate the robustness of two networks: the 68-bus and the 118-bus IEEE systems. Finally, some concluding remarks are provided in Sect. 4.3.

4.1   Quantification of the Network Robustness

We focus on the topological aspects of the power systems to quantify its robustness. Robustness of the power grid is related to the topological firmness of the interconnection among substations through of the transmission system. The topological firmness is also related to the relative importance of the buses inside the network because the most important nodes play an important role in the connectivity of the whole network. The power system reliability depends significantly of the network robustness because if the interconnection level among substations is low then the system integrity can be impacted by some disturbances. In other words, a strong or robust network means that there are enough connecting paths that increase the system reliability in various aspects including the interchange of resources.

The relative importance of each node may be measured using the topological characterization for power system networks developed in Chap. ??. The worth of every node is determined for the quality of the nodes which it is connected. If a particular node is connected to various important nodes then this node plays a meaningful role inside of network in terms of connectivity. If we assume there is a vector w of positive values w_i indicating the importance of the node i then the importance of the node i is proportional to the sum of the importance measures of the all nodes connected to it.

$$w_i = \frac{1}{\lambda} \sum_{j=1}^{N} a_{ij} w_j \tag{4.1}$$

Where a_{ij} is an entry (i, j) of the Laplacian matrix $\mathcal{L}$. This entry indicates if the node pair (i, j) is connected or not. This can be rewritten as,

$$\mathcal{L} = \lambda \tag{4.2}$$

There will be many different eigenvalues and eigenvectors by pairs of the matrix that yields this condition. However, the requirement over of positives entries implies that the solution of (4.2) is unique by the Perron–Frobenius theorem [12]. The statement of the theorem is as follows:

Perron–Frobenius theorem: if $\mathcal{L}$ is a $n \times n$ non-negative irreducible matrix, then there is a largest eigenvalue λ_{PF} such that,

(i) λ_{PF} is positive (ii) there exist an unique eigenvector $\mathbf{w} > 0$ such that $\mathbf{Lw} = \lambda_{PF}\mathbf{w}$ (iii) λ_{PF} for any eigenvalue λ_{PF}

1. is positive

   a. Livelihood and survival mobility are oftentimes coutcomes of uneven socioeconomic development.
   b. Livelihood and survival mobility are oftentimes coutcomes of uneven socioeconomic development.

2. Livelihood and survival mobility are oftentimes coutcomes of uneven socioeconomic development.

Thus, this chapter proposes to use directly the eigenvalue as a measure of robustness. This eigenvalue is bounded by the maximum degree that a network can reach when every node is connected,

$$\lambda_{PF} \leq dmax \tag{4.3}$$

Fig. 4.1 Fully connected
5-bus network

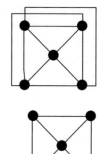

Fig. 4.2 5-bus network with
a robustness index of 81%

A network of n nodes reaches his maximum robustness if every node is connected with $n - 1$ nodes. In other words, the eigenvalue λ_{PF} reaches its maximum theoretical value that is equal to $n - 1$. Then, the robustness index γ is given by,

$$\gamma = \frac{\lambda_{PF}}{n - 1} \tag{4.4}$$

Each network has some intrinsic properties characterizing its structure. For a network $\mathcal{G} = \{\mathcal{N}, \mathcal{L}\}$ the set of lines $\mathcal{L}$ connect the set of nodes $\mathcal{N}$ in of a particular manner, the structural properties such as the robustness can change radically by modifications of its structure, specifically by the addition or removal of the links. The robustness index allows suitable quantification of these topological changes. For instance, the 5-bus system in the Fig. 4.1 is a fully connected network; the corresponding Perron-Frobenious eigenvalue is equal to $\lambda_{PF} = 4$ and; therefore, the network robustness index is equal to 100%. We can observe that the diameter of this network is equal to 1 since one link can be used to connect any pair of nodes.

Now, to illustrate how the robustness index changes if the structure is modified, we remove two links of the previous network as shown in Fig. 4.2

The robustness index for the modified 5-bus network is equal to 81% and the diameter is equal to 2. If we again remove three links, as shown in Fig. 4.3 we obtain a network robustness of 46% and the diameter is equal to 3. This network depends strongly of some links to maintain its connectivity; for instance, just one link is connecting one node to other. If this link is removed, the network will be disconnected.

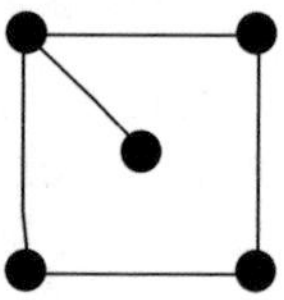

Fig. 4.3 5-bus network with
a robustness of 46%

The power system networks are not highly meshed in comparison with other networks because the construction of transmission lines involves very high costs. Thus, the power system networks exhibit generally high ratio of nodes and links and, therefore, they are characterized by a weak connectivity and; therefore, the fairly low robustness index. In other words, the diameter of the network is large, which indicates a network not very well connected. In the application of this metric to power systems, we consider that 50% network robustness is good enough if the ratio of nodes and links is large, the robustness index could be insensitive to small changes in the network topology. This document computes explicitly the robustness of the power systems networks using subnetworks to obtain a measure that describes well the topological firmness.

4.2 Illustrative Examples: 68-Bus and 118-Bus IEEE Test Systems

The implementation of the proposed approach is straightforward and we will illustrate its application to two networks: the 68-bus and the 118-bus IEEE systems. In the connected 68-bus system, Fig. 4.4 we will compute the robustness index for

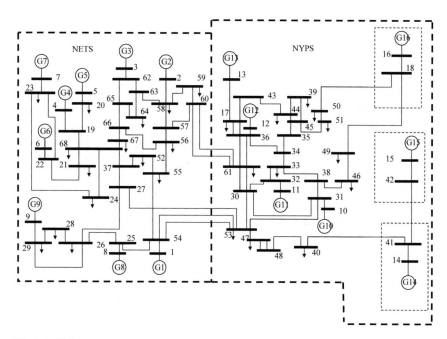

Fig. 4.4 68-bus test system

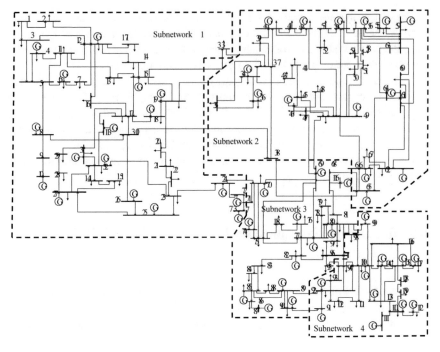

Fig. 4.5 IEEE 118-bus system and four subnetwork

subnetworks representing the New England Test System (NETS) and the New York Power System (NYPS).

The corresponding Laplacian and Adjacency matrices for the NETS and NYPS are computed and the largest eigenvalues (Perron-Frobenious eigenvalue) are $\lambda_{NETS} = 6.4$ and $\lambda_{NYPS} = 5.8$ respectively. Each power subsystem exhibits similar network robustness, the index for NETS is $\gamma_{NETS} = 18.3\%$ and $\gamma_{NYPS} = 18.6\%$ for NYPS, which indicate that NETS and NYPS are weakly connected. These subnetworks should be reinforced to reach 50% of the robustness value. Next, we will examine the IEEE 118-bus system and quantify its robustness. The robustness index is computed for four subnetworks shown in Fig. 4.5. The Adjacency matrix and the corresponding largest eigenvalue λ_{PF} associated with each subnetwork are calculated. Table 4.1 summarizes these results.

We can observe that each subnetwork exhibits different value of the robustness index, which indicates that some areas are more strongly connected than the others. For instance, the subnetwork 1, which has the lowest index, requires structural reinforcement to increase its robustness through the expansion of the transmission system. The subnetworks 2 and 3 also require structural reinforcement to reach at least 50% of the robustness value.

Table 4.1 Robustness index for EEE 118-buses system

	Largest eigenvalue	Network robustness
Subnetwork 1	$\lambda_{sub1} = 9.29$	$\gamma_{sub1} = 21.3$
Subnetwork 2	$\lambda_{sub2} = 8.31$	$\gamma_{sub1} = 28.3$
Subnetwork 3	$\lambda_{sub3} = 7.76$	$\gamma_{sub1} = 32.3$
Subnetwork 4	$\lambda_{sub4} = 9.27$	$\gamma_{sub1} = 51.5$

4.3 Concluding Remarks

This chapter develops a graph-theoretic-algebraic approach to quantify the robustness of the power system networks and to provide insight into their topological firmness. The new index can be very useful in the planning environment for making decisions about the reinforcement of the power system structure. When the topology of the system changes, for maintenance of lines, outages of lines, etc., the index could be used to quantify whether the network robustness is improved or not. The quantification of the network robustness will allow the transmission system owners and regulators to obtain useful information when planning an expansion of the power grid. The network structure plays a critical role in the power system performance but the structural properties of the grids have not been given enough attention in the power system studies. Calculation of the robustness index could help to remedy this situation.

References

1. Tobon AG, Chamorro HR, Gonzalez-Longatt F, Sood VK (2019) Reliability assessment in transmission considering intermittent energy resources. In: 2019 IEEE 10th latin American symposium on circuits systems (LASCAS), pp 193–196
2. Chamorro HR, Sanchez AC, Øverjordet A, Jimenez F, Gonzalez-Longatt F, Sood VK (2017) Distributed synthetic inertia control in power systems. In: 2017 international conference on energy and environment (CIEM), pp 74–78
3. Courant R, Hilbert D, Methods of Mathematical Physics, Interscience, New York, Vol. 1, 1953
4. Newman MEJ, The structure and function of complex networks 45(2):167–256. http://dx.doi.org/10.1137/S003614450342480
5. Raak F, Susuki Y, Hikihara T, Chamorro HR, Ghandhari M, Partitioning power grids via nonlinear Koopman mode analysis. In: ISGT 2014, pp 1–5. http://dx.doi.org/10.1109/ISGT.2014.6816374
6. Torres A, Anders G (2009) Spectral graph theory and network dependability. Fourth international conference on dependability of computer systems 2009:356–363
7. Dekker AH, Colbert BD, Network robustness and graph topology. In: Proceedings of the 27th Australasian conference on computer science - volume 26, ACSC '04. Australian Computer Society, Inc., pp 359–368

8. Van Mieghem P (2005) Robustness of large networks. In: 2005 IEEE international conference on systems, man and cybernetics, vol. 3, pp 2372–2377
9. Jamakovic A, Uhlig S, On the relationship between the algebraic connectivity and graph's robustness to node and link failures. In: 2007 Next generation internet networks, pp 96–102. http://dx.doi.org/10.1109/NGI.2007.371203
10. Sasson AM, Ehrmann ST, Lynch P, Van Slyck LS, Automatic power system network topology determination PAS-92(2):610–618. http://dx.doi.org/10.1109/TPAS.1973.293764
11. Prais M, Bose A, A topology processor that tracks network modifications 3(3):992–998. http://dx.doi.org/10.1109/59.14552
12. Perron O, Zur theorie der matrices 64:248–263. https://eudml.org/doc/158317

Chapter 5
Security Strategies Applications

Abstract Modern power systems are highly interconnected networks, which are operated closer to their security limits. In the last decade, power systems have been under stressed conditions as a result of new economic requirements and growth in generation and demand without corresponding expansion of the transmission networks.

Recently, several blackouts have occurred around the world, revealing an evident weakness of the power systems. Many factors contribute to the catastrophic failures in the power systems, including weak connections, unexpected events, natural failures, external attacks, operator failures, and hidden failures in the protection systems.

The latest major disturbances reported in the literature showed that a very large part of a power system can be affected by a widespread event, including places that are electrically far away from the place where the fault occurred. The cascading failures can cause a major blackout when they are not appropriately managed by the power system operators. Table 5.1 present the loss of load and the duration of the blackouts that happened in the last decade, based on the corresponding reports [1–5].

In general, the sequences of events in the major blackouts follow a common process. In the first stage, the cascading process can advance at a relatively slow speed [6], as happened, for example, during the U.S and Canada 2003 blackout [1]. In the second stage, the successive outages of the transmission lines and generation station can spread quickly and uncontrollably [7]. In summary, blackouts often start with a local failure in the electric network and turn into a major event due to a succession of diverse and rare events.

In consequence, power systems require strategies to mitigate catastrophic failures. A feasible strategy should protect a major part of the network and maintain the operational conditions and requirements of electricity supply. Control strategies play an important role in managing the system behavior. A well-designed control strategy in case of a major disturbance could involve first separation of the network into islands and then restoration of the system. This chapter addresses the first aspect of the problem.

© The Author(s), under exclusive license to Springer Nature Switzerland AG 2021   27
R. Moreno Chuquen et al., *Graph Theory Applications to Deregulated Power Systems*,
SpringerBriefs in Electrical and Computer Engineering,
https://doi.org/10.1007/978-3-030-57589-2_5

Table 5.1 Robustness index for EEE 118-buses system

Blackout	Loss of load (MW)	Duration
US-Canada, 2003	61, 800	72 h
UK, 2003	724	37 min
Sweden, 2003	6550	8 h
Italy, 2003	24, 000	12 h
Colombia, 2007	7, 083	4.5 h
Brazil, 2009	28, 000	8 h

5.1 Hierarchical Islanding of Power System as a Security Strategy

This chapter proposes the application of a security strategy based on the creation of islands based on the identified cut-set of transmission lines. We call this strategy as Hierarchical Islanding; this security strategy develops a procedure to separate power systems in order to mitigate widespread disturbances and cascading effects during catastrophic failures that compromise the security of the entire power system. The islanded system represents a temporal operational condition, allowing the power system operators to take corrective actions and to initiate a restoration process of the area where the failures has occurred.

The method proposed in this chapter, based on a spectral graph analysis, allows the identification of the islands with minimum transfer admittance between them. We use directly a modified laplacian matrix based on the approximately admittance matrix as it's was developed in the Chap. 2 to identify subnetworks.

This chapter presents the formulation and general algorithm to develop a procedure for hierarchical islanding of power systems based on spectral graph theory. The power system is made up of different resources such as power generation centers, transmission lines, transformers, demand sides, and so on, and these resources are interconnected together to perform the function of supplying electric energy, economically, reliably and safely around the clock. The secure operation of power systems should be maintained into secure states under different grid impactive events. Such operational and nonoperational grid impactive events occur infrequently, but when they do, they reveal the weakness of the system and the need for security strategies to counteract impactive events such as cascading failures.

Hierarchical Islanding of Power System as a Security Strategy develops a procedure to separate power systems in order to mitigate widespread disturbances and cascading effects during catastrophic failures that compromise the security of the entire power system. The islanded system represents a temporal operational condition, allowing the power system operators to take corrective actions and to initiate a restoration process of the area where the failures has occurred. The method proposed in this project, based on a spectral graph analysis, allows for the identification of the islands with minimum transfer admittance between them.

The power system operators must use effective strategies to maintain secure system operations under all the contingency cases that are considered. The design of such strategies must provide robust response actions, which ensure that static security is achieved by restoring the power system security to a normal operating state. The hierarchical islanding of power systems allows the construction of such strategies by explicitly exploiting the properties of the topological structure. Specifically, the proposed strategies make use of the identified islands that are created by the removal of one or more tie lines to reduce the impact of grid-impactive events such as cascading outages. The switching actions over selected transmission lines are operational resources to be dispatched by the power system operator to ensure the security under various grid-impactive events to form islands and separate a sick part of the power grid. The knowledge of the best way to separate the power system into several islands offers relevant information about the islanding capability of the system. The islanding as an operational resource can be used to divide the network into islands upon successive instructions from the user [8].

A general algorithm to develop a procedure for hierarchical islanding of power systems is described in this section. The hierarchical islanding (HI) procedure will read the network data in order to construct the topological model of power system and upon successive instructions from the user will divide the power system into two or more islands using spectral graph analysis. Since the interest is in the identification of islands to mitigate cascading outages, the HI exploits the electrical distance concept to find islands which are electrically close among themselves, while buses of different islands are electrically distant from each other. The general algorithm is presented in Fig. 5.1 and explained as follows:

The islanding request process performs the request by the user to create two islands. The first request will divide the entire network into two islands, each of which can be subsequently divided further. If there is a request to divide the network or any subnetwork into two islands then a spectral analysis of the system should be executed. The spectral analysis module involves spectral graph analysis of the power system, which allows the deduction of principal properties and the structure of a graph from its graph spectrum. The spectrum of a graph is the set of eigenvalues associated with the various matrices representing the graph. In a graph with n vertices, there are n eigenvalues. The analysis of the graph eigenvalues and eigenvectors provides an important insight into understanding the graphs in power systems. The spectral method employs global information about the graph by computing a separator from eigenvector components. The procedure is carried out by the spectral analysis of the Laplacian matrix.

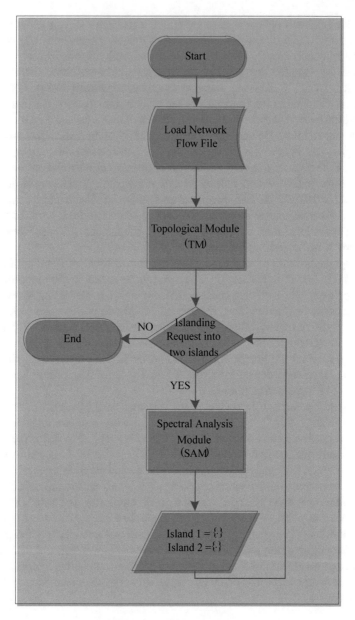

Fig. 5.1 General algorithm to develop Hierarchical Islanding

5.2   Illustrating the Hierarchical Islanding Procedures

5.2.1   A Medium Size Network Example

This section illustrates the application of the Hierarchical Islanding procedure to standard electrical system in order to maintain secure system operations under contingency cases. The standard 68-bus test system is used to show the HI procedure performance. This test system is a reduced order equivalent of the interconnection between the New England Test System (NETS) and the New York Power System (NYPS). Figure 5.2 shows the standard 68-bus test system diagram.

The identification of two islands is obtained when the Hierarchical Algorithm is executed. Figure 5.3 shows graphically the buses belonging to each island and the Table 5.2 summarizes this information. This separation is interesting because the Island 1 and Island 2 matches up with the connection between NETS and NYPS, as was expected.

If there is a new request to separate either Island 1 or Island 2 into two islands, to reduce the impact of any cascading outage, then a new separation is executed. Figure 5.4 shows the separation of the Island 1 into two islands and Fig. 5.5 the separation the Island 2 into two islands.

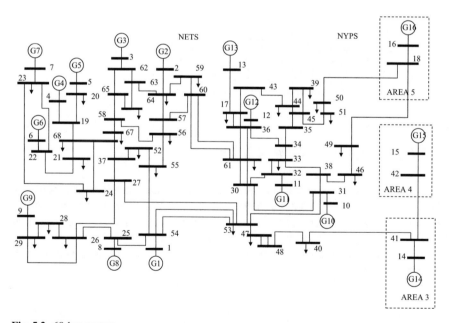

Fig. 5.2   68-bus system

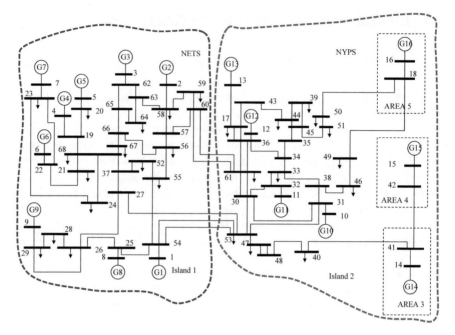

Fig. 5.3 Separation of 68-bus test system into two Islands

Table 5.2 Two subnetworks for the 68-bus system

Island 1	1, 2, 3, 4, 5, 6, 7, 8, 9, 19, 20, 21, 22, 23, 24, 25, 26, 27, 28, 29, 37, 52, 54, 55, 56, 57, 58, 59, 60, 62, 63, 64, 65, 66, 67, 68,
Island 2	10, 11, 12, 13, 14, 15, 16, 17, 18, 30, 31, 32, 33, 34, 35, 36, 38, 39, 40, 41, 42, 43, 44, 45, 46, 4748, 49, 50, 51, 53, 61

5.2.2 A Large Size Network Example

This section illustrates the application of the Hierarchical Islanding procedure to a large scale test system in order to maintain secure system operations under contingency cases. The standard 118-bus test system is used to show the HI procedure performance. Figure 5.6 shows the 118-bus test system diagram.

When a need arises, the HI procedure is executed to identify two islands with minimum transfer admittance between them. The islands are shown in Fig. 5.7.

The Hierarchical Islanding procedure yields the division of buses as shown in Table 5.3.

The HI procedure goal is to mitigate the impact of grid impactive events such as cascading outages. A second request could be made by the user to divide either of the islands into two parts. In order to reduce the impact of grid impactive events, then,

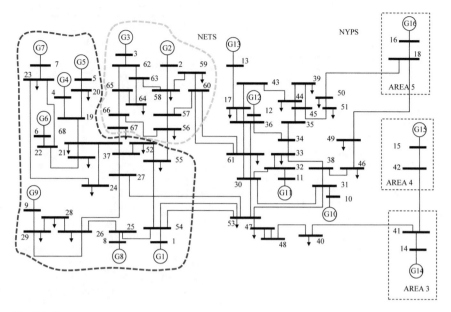

Fig. 5.4 Second request to divide the Island 1 into two Islands

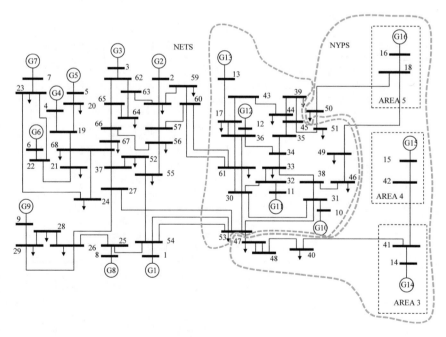

Fig. 5.5 Second request to divide the Island 2 into two Islands

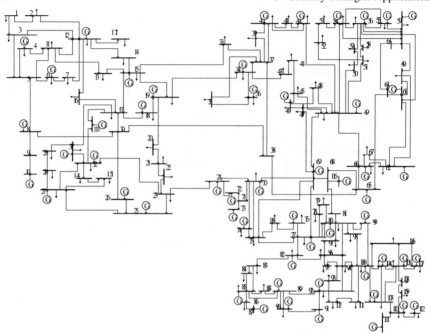

Fig. 5.6   118-bus test system

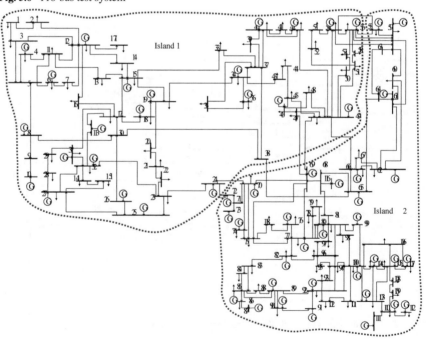

Fig. 5.7   Separation of the 118-bus test system into two Islands

Table 5.3 Division of 118-bus system into two Islands

Island 1	1, 2, 3, 4, 5, 6, 7, 8, 9, 10, 11, 12, 13, 14, 15, 16, 17, 18, 19, 20, 21, , 22, 23, 24, 25, 26, 27, 28, 29, 30, 31, 32, 33, 34, 35, 36, 37, 38, 39, 40, 41, 42, 43, 44, 45, 46, 47, 48, 49, 50, 51, 52, 53, 54, 55, 56, 57, 58, 72, 113, 114, 115, 117
Island 2	55, 59, 60, 61, 62, 63, 64, 65, 66, 67, 68, 69, 70, 71, 72, 73, 74, 75, 76, 77, 78, 79, 80, 81, 82, 83, 84, 85, 86, 87, 88, 89, 90, 91, 92, 93, 94, 95, 96, 97, 98, 99, 100, 101, 102, 103, 104, 105, 106, 107, 108, 109, 110, 111, 112, 116, 118

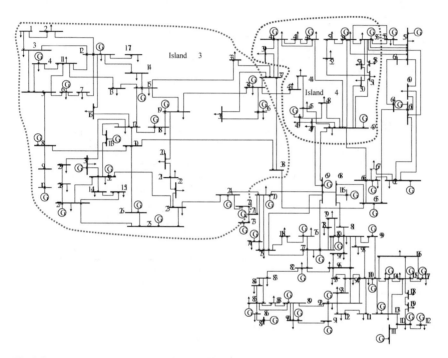

Fig. 5.8 Separation of the Island 1 into two Islands

a new division of the system could be computed. Figure 5.8 shows the separation of Island 1 into two islands labeled Island 3 and Island 4, respectively.

Thus, if a grid impactive event is detected inside of the Island 3, the impact could be reduced if this island is separated from the entire network. In the same way, the Island 4 could be kept apart from the entire network reducing the impact of a cascading event. Table 5.4 shows the groups of buses belonging to Island 3 and Island 4 when the HI procedure is executed.

On the other hand, if an alarm is detected inside of the Island 2, then a request to divide the Island 2 into two islands allow the mitigation of a cascading outage. Figure 5.9 shows schematically the separation of Island 2 into two islands labeled Island 5 and Island 6, respectively. Table 5.5 shows the groups of buses belonging to Island 5 and Island 6 when the HI procedure is executed.

Table 5.4 Buses belonging to the Islands 3 and 4 for the 118-bus system

Island 3	1, 2, 3, 4, 5, 6, 7, 8, 9, 10, 11, 12, 13, 14, 15, 16, 17, 18, 19, 20, 21, 22, 23, 24, 25, 26, 27, 28, 29, 30, 31, 32, 33, 34, 35, 36, 37, 38, 72, 113, 114, 115, 117
Island 4	39, 40, 41, 42, 43, 44, 45, 46, 47, 48, 49, 50, 51, 52, 53, 54, 56, 57, 58

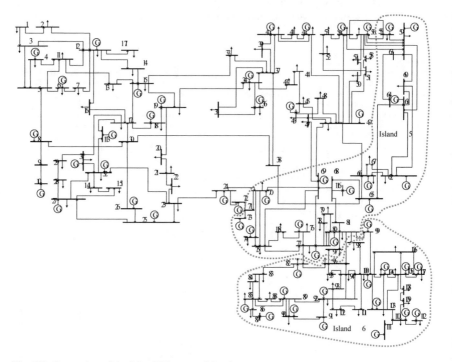

Fig. 5.9 Separation of the Island 2 into two Islands

Table 5.5 Buses belonging to the Islands 5 and 6 for the 118-bus system

Island 5	55, 59, 60, 61, 62, 63, 64, 65, 66, 67, 68, 69, 70, 71, 73, 74, 75, 76, 77, 78, 79, 80, 81, 97, 116, 1118
Island 6	82, 83, 84, 85, 86, 87, 88, 89, 90, 91, 92, 93, 94, 95, 96, 98, 99, 100, 101, 102, 103, 104, 105, 106, 107, 108, 109, 110, 111, 112

5.2.3 Concluding Remarks About Hierarchical Islanding

The formulation and algorithm of Hierarchical Islanding of power systems based on the spectral graph theory has been developed as a security strategy to mitigate grid impactive events such as cascading outages. The security strategy based on the successive identification of the islands with minimum transfer admittance between them is an effective operational resource to manage the power system under contingencies cases and provides robust response actions. A procedure presented in the report that divides the power system into several islands offers relevant information about the islanding capability of the system in order to perform security studies for real-time and offline environments. For instance, the identification of islands could be effectively used in the studies of the restoration processes of power systems following a multiple contingency. An efficient algorithm has been developed to compute the second eigenvector of the Laplacian matrix. This algorithm uses a Lanczos method exploiting the sparsity of the Laplacian matrix. The algorithm is highly adaptable because is based on two independent modules (TM and SAM) which allow the implementation of new functionalities when needed.

The presented methodology allows finding different scenarios of system partitioning with a minimum number of transmission lines belonging to a cut set. The methodology could be applied to a large system with a fast computation time when the Lanczos method is used, because the computation time grows linearly with the number of transmission lines.

The hierarchical separation discussed in this chapter is one of the mechanisms in the defensive strategies against critical contingencies that compromise the security of entire power systems.

5.3   Quantification of Grid-Impactive Events

5.3.1   Identification of Critical Transmission Lines

The identification of critical transmission lines in power systems is prerequisite for security assessment studies. Based on our network topological approach we identify the minimum set of transmission lines that interconnect subnetworks into an interconnected power network. The critical transmission lines are directed related with the capability of an interconnected network to transfer power trough of the network. The outage of one or more critical transmission lines for different disturbances could reduce dramatically the connectivity of the system and consequently the capability to transfer power. We propose the construction of a graph-model to study the connectivity and determine the transmission lines that keep the connectivity of a power network. We also study the impact of the outage of various transmission lines over the set of critical transmission lines in order to determine overflows over the critical transmission lines that could provoked overload and a trip out action.

In this section, we illustrate how to identify critical transmission lines using indirectly the results obtained from the Chap. 3 and also the usage of this information to perform security studies.

First, we use the spectral information deduced from the Laplacian matrix of the 68-bus system to identify the set of critical transmission lines that interconnect two subnetworks. As it expected, we find that the set of critical transmission lines are

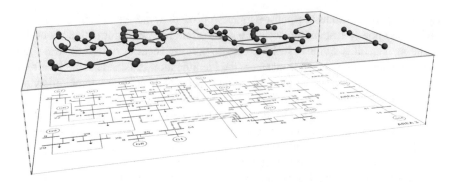

Fig. 5.10 Critical transmission lines connecting NETS and NYPS in 68-bus system

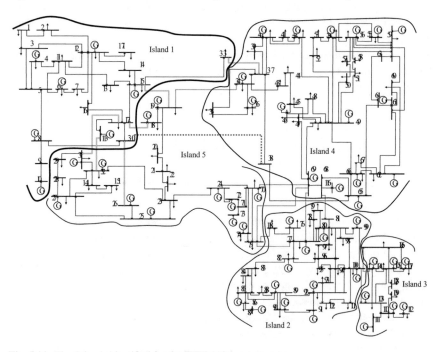

Fig. 5.11 Five Islands identified for the IEEE 118-bus test system

because these lines interconnect the subnetworks NETS and NYPS, as it's schematically displayed in Fig. 5.10

We now examine the impact of the outage of one transmission line connecting one subnetwork with the whole network. We use the so-called Power Transfer Distribution Factors (PTDFs) [9] to evaluate the impacts of line outages on the non-outaged lines' flows in terms of the so-called line outage distribution factors (LODFs). Based on the identification of the tie transmission lines connecting various subnetworks, we use directly the LODFs values to provide the fractions of the pre-outage flow

Table 5.6 LODFs for the 118-bus system

Line	Fraction of the pre-outage flow of the outaged line
{33, 37}	0.36
{15, 19}	0.14
{17, 18}	0.21
{30, 26}	0.21
{17, 31}	0.03
{113, 32}	0.03

on the outaged line that are redistributed to the non-outage lines in the post-outage network. Figure 5.11 shows five islands. Particularly, the island 1 is connected with the whole network through of the transmission lines: {33, 37}, {15, 19}, {17, 18}, {30, 38}, {30, 26}, {113, 32}, {17, 31}. Table 5.6 summarizes particularly the impact of the outage of the transmission line {30, 38} over the others transmission lines that connect the island 1 with the whole network.

We also analyze a set of critical transmission lines in the IEEE 39-bus system. Using the spectral graph results deduced before to identify two subnetworks we found

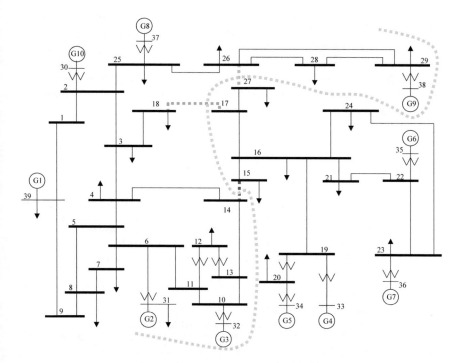

Fig. 5.12 Two subnetworks in the IEEE 39-bus system

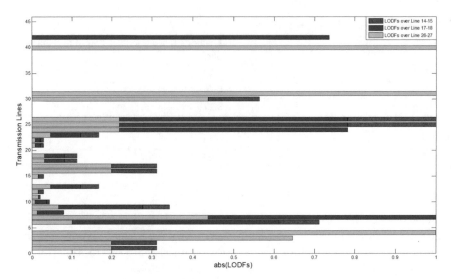

Fig. 5.13 LODFs of all lines when the lines are tripped out

that the set of transmission lines {26, 27}, {18, 17}, {14, 15} connect two subnetworks through of a minimal number of lines as is showed in Fig. 5.12.

Now, we perform a security analysis study to quantify the impact of the outage of the critical lines over all others transmission lines. Specifically, we compute the LODFs of all lines when the lines {26, 27}, {18, 17}, {14, 15} are tripped out. Figure 5.11 shows the power fraction that is redistributed over each line. The red bar indicates the power fraction of the line {14, 15} that is redistributed over all lines, the blue bar indicates the power fraction of the line {18, 17} and the green one indicates the power fraction of the line {26, 27} (Fig. 5.13).

References

1. Andersson G, Donalek P, Farmer R, Hatziargyriou N, Kamwa I, Kundur P, Martins N, Paserba J, Pourbeik P, Sanchez-Gasca J, Schulz R, Stankovic A, Taylor C, Vittal V (2005) Causes of the 2003 major grid blackouts in north America and Europe, and recommended means to improve system dynamic performance. IEEE Trans Power Syst 20(4):1922–1928
2. Bialek JW (2005) Blackouts in the US/Canada and continental Europe in 2003: Is liberalisation to blame? IEEE Russia Power Tech 2005:1–7
3. Berizzi A (2004) The Italian 2003 blackout. In: IEEE power engineering society general meeting, 2004, vol 2, pp 1673–1679
4. Vournas CD, Nikolaidis VC, Tassoulis AA (2006) Postmortem analysis and data validation in the wake of the 2004 Athens blackout. IEEE Trans Power Syst 21(3):1331–1339
5. Ruiz CA, Orrego NJ, Gutierrez JF (2008) The colombian 2007 black out. In: 2008 IEEE/PES transmission and distribution conference and exposition: Latin America, pp 1–5. http://dx.doi.org/10.1109/TDC-LA.2008.4641845

6. Makarov Y, Reshetov V, Stroev A, Voropai I, Blackout prevention in the United States, Europe, and Russia 93(11):1942–1955. http://dx.doi.org/10.1109/JPROC.2005.857486
7. Pourbeik P, Kundur P, Taylor C, The anatomy of a power grid blackout - Root causes and dynamics of recent major blackouts 4(5):22–29. http://dx.doi.org/10.1109/MPAE.2006.1687814
8. Chamorro HR, Ordonez CA, Peng JC, Ghandhari M (2016) Non-synchronous generation impact on power systems coherency. IET Gener, Trans Distrib 10(10):2443–2453
9. Power generation, operation, and control, 3rd edn. Wiley, New Jersey

Chapter 6
Cyber Physical Systems Security for the Smart Grid

Abstract The integration of Information Technology (IT) systems (computations and communication—the cyber world) with sensor and actuation data (the physical world), can introduce new, and fundamentally different approaches to security research in the growing field of Cyber-Physical Systems (CPS), when compared to other purely-cyber systems [1]. Earlier work [2–5] have shown that because of the automation and real-time requirements of many control actions, traditional security mechanism are not enough for protecting CPS, and we require resilient control and estimation algorithms for true CPS defense-in-depth [6]. In this chapter we outline how attacks and resilient mechanisms can affect and defend power grid operations.

6.1 Energy Management Systems

To characterize the CPS security of the power grid, we need to understand how IT is used in the control centers of the power grid to collect sensor data, estimate the state of the power grid, and issue control commands and pricing signals to the market [7].

One of the most important components in a control center is the Energy Management System (EMS). The EMS is responsible for many operational tasks. It includes the Network Topology Processor (NTP), state estimation, the market process, and control actions for transmission automation, such as remote tap adjustment for transformers [8].

The role of NTP, and state estimation is to collect data from sensors in the field, and give an accurate view (topology and electricity flows). If the data collected is incorrect, operators will get an erroneous view of the system and all management functions of the control center will be affected, (including market computations and control actions). This is the reason why a lot of recent work has focused on deception attacks (also known as false-data injection), where a compromised sensor sends malicious data back to the EMS.

© The Author(s), under exclusive license to Springer Nature Switzerland AG 2021   43
R. Moreno Chuquen et al., *Graph Theory Applications to Deregulated Power Systems*,
SpringerBriefs in Electrical and Computer Engineering,
https://doi.org/10.1007/978-3-030-57589-2_6

In this chapter we survey recent work on CPS security for power systems and present the work in a unified view by showing how all previous attacks are parts of the EMS. We find some limitations with previous work and discuss open problems and new research challenges of parts of the Ems that have not been considered in previous work.

6.2   Cyber Physical Systems Security for the Smart Grid

The state estimation problem in power systems originates from the need of power engineers to estimate the phase angles $x \in$ from the measured power flow $z \in$ in the transmission grid. It is know that the measured power flow $z = h(x) + e$ is a nonlinear noisy measurement of the state of the system x from and an unknown quantity e called the measurement error. To estimate x from this set of equations engineers make usually two simplifications: (1) e is assumed to be a Gaussian noise vector with zero mean an covariance matrix W and (2) the equation is approximated by the linear equation,

$$z = H(x) + e \tag{6.1}$$

where H is a matrix. Estimating x from these equations is achieved by computing the Minimum Mean Square Error (MMSE) estimate:

$$\hat{x} = \left(H^T W H\right)^{-1} H^T W z \tag{6.2}$$

Because the transmission system is composed of thousands of sensors (i.e., z is a vector of thousands of scalars) and not all sensors are reliable, power engineers have devised a set of test to detect bad measurements. The tests are based on the following test:

$$\left\| z - H\hat{x} \right\| > \tau \tag{6.3}$$

that is, if the measurement and the estimated measurement are greater than a threshold, then the test decides that there are some faulty sensors in the system. If the test has a value lower or equal to then the test concludes that all measurement are correct.

Liu et al. [9] introduced attacks against the integrity of the state estimation algorithms in the power grid by showing that there attacks where a compromised sensor can send a false measurement reading and yet the bad data detection test will not detect this attack. In particular, they show how by selecting an attack signals $z_\alpha = z + a$ where $a = H_c$ for any vector c creates a successful attack. Then they analyzed how attackers can craft these attacks when they have different resources (limited access to meters or limited ability to compromise meters) and different objectives (random attacks or specific error in the estimate). While attack in larger systems are difficult to create (in an IEEE 3000 bus system the attacker needs to compromise more than 900 m) and may have limited negative effects (the injected error might not be too large), the fact that attackers can manipulate the view of one of our critical infrastructures is a worrisome fact.

Some follow up work has discussed extensions on how to better protect the power grid to these attacks. Some preliminary results in this area of research include the work of Dan and Sandberg [10], who considers a defender that can secure individual measurements by, for example, replacing an existing meter to a meter with better security mechanism such as tamper resistance of hardware security support. Their goal is to protect the system under a limited budget and to that end they formulate the problem as identifying the best measurements to protect (they assume the attacker cannot compromise these sensors) in order to minimize the impact of attacks. The mathematical problem they consider is a combinatorial optimization, so this problem is intractable for large systems. The main contribution of this work is to exploit the structure of the power system matrices to make the optimization problem efficient. Kosut et al. [11] also extend the basic false data injection attack to consider attackers trying to maximize the error introduced in the estimate, and defenders with a detection algorithm that attempts to detect false data injection attacks. Their new detection algorithm performs better than the traditional bad data detection algorithms (since these algorithms were designed for detecting faults, not network attacks). Their detection algorithm is based on the generalized likelihood ration test, which is not a tractable problem to solve.

6.3   Network Topology Processor

Each breaker in the transmission system has a sensor reporting if it is open or closed. This information is sent to the NTP to construct the topological model of the system [12]. This topological model is used for the state estimation of the system. If the topology is wrong, the state estimation algorithm will also produce erroneous results.

As far as we know, no previous work has studied the false-data injection problem against the NTP.

6.4   Electricity Markets

The goal of the electricity market process in the control center is to deliver market prices such as LMPs. Particularly, LMPs are computed at each load and at each generation bus when the transmission system is congested (which is the default state) to determine how much will utilities pay the system operator, and how much will the system operator pay to the generators. LMPs are traditionally, depends of the system, computed every 5–10 min, but there is recent work (e.g., New York Power System) for computing LMPs in real-time.

Quantifying the cost of security incidents is one of the most difficulty problems in computer security because it is hard to quantify the value of information. However, by analyzing attacks against the electricity market, we can quantify the effects of these attacks by leveraging the economic metric used to measure the efficiency of the system [13].

As we mentioned before, it the state estimation is incorrect, all management functions of the control center are affected, including the market operations. Xie et al. [14] studied how false data injections attacks can be used to defraud deregulated electricity markets by modifying LMPs. They consider the case where attackers can manipulate prices while bring undetected by the system operator.

In all other attacks considered in this paper, the attacker can be implicitly assumed to be malicious entity that tries to destabilize the system or reduce the social welfare. On the other hand, the work of Xie et al. considers a selfish attacker instead of a malicious attacker. This change in the motivation of the attacker makes it difficult to understand which party will have the long-term motivation to launch these types of attacks. Utilities, generators, and system operators are large, highly regulated companies who have higher incentives to remain in business than to launch an attack that put their company in jeopardy (in case it is discovered).

So far, all attackers presented in this chapter were based on false-data injection attacks against the sensor data used for state estimation. Gross et al. consider a new type of attacks by studying the integrity of the control signals (as opposed to the integrity of the sensor signals). In particular, they study how malicious control signals sent to circuit breakers (directing them to remove transmission lines from the system) affect the social welfare metric of the market system.

6.5   Transmission Automation

In addition to the control signals sent to circuit breakers, as considered by Gross et al. there are many other control signals that can be falsified by an attacker, in particular, given that the smart grid is introducing the capability of more distributed, automatic control.

The Flexible Alternate Current Transmission System (FACTS) includes many automatic electronic devices such as Static Voltage Compensators (SVC), uses reactive power to improve the voltage profile of the system. Similarly, the Thyristor Controlled Series Compensator (TCSC) is a control device in series with a transmission line which can be used to modify its impedance to control the current going through these lines. Taxonomy of attacks against FACTS devices was presented by Phillips et al. [15] and an implementation of some attacks with false status reports and control actions showed unnecessary VAR compensation and unstable operation of the system [16].

Other control signal that can sent remotely include tap adjustment for smart transformers (used to increase or decrease slightly the voltage on each side of transformer), and the Automatic Generator Control (AGC) signal (which is used to set the voltage of generators). Robust attack policies have been studied for AGC signals [17],

[18], and attacks have shown that if you modify the frequency and tie-lien flow measurements, the system can be driven to abnormal operating values [16].

6.6 Defense Mechanisms

In addition to traditional IT security mechanism for prevention (authentication, encryption, firewalls) and detection (intrusion detection systems, forensics) we need new CPS security mechanisms [19]. There are several CPS planning and defense mechanism that can leverage knowledge of the attacks presented in this chapter. The first is risk assessment: given a fixed budget, where should one allocate this budget to minimize my potential physical damages?

A second mechanism is bad data detection mechanisms. These mechanisms should not assume random, independent failures, but consider detection of sophisticated attackers. Interestingly enough, most previous work has focused on attacks, but very few have proposed novel attack-detection mechanisms [11]. One particular open problem is to propose bad topology detection mechanisms.

Replacing sensed data with false data (a deception attack) is a very generic attack that can be extended to any smart grid application (as all of them are based on correct sensor measurements). It is important to develop intrusion detection mechanisms or reputation management system for smart grid application where not all received data can be trusted.

The defense third mechanism is to introduce resiliency (or survivability) of the system to attacks. A promising direction is to design the topology of the power distribution network to withstand malicious commands to circuit breakers trying to change and disconnect the network.

6.7 Concluding Remarks and Future Work

CPS security is a growing field critical for the vision of a survivable power grid that can withstand attacks and reconfigure or adapt to mitigate adverse effects. Work on fault tolerance and reliability of control system is not enough, because these mechanisms generally assume independent and uncorrelated failures; however, cyberattacks will exploit vulnerabilities in a coordinated and correlated fashion. The most basic example is the work of false-data injection attacks, where it is shown that traditional safety and fault-detection mechanism currently available in the power grid cannot detect incorrect sensor data when a malicious attacker is the source of these errors. Therefore instead of relying solely on fault-detection algorithms to protect control algorithms in the power grid, we need to develop new attacks-detection algorithms focusing on identifying malicious data in sensor actuation devices in the power grid.

References

1. Chow JH, Vanfretti L, Armenia A, Ghiocel S, Sarawgi S, Bhatt N, Bertagnolli D, Shukla M, Luo X, Ellis D, Fan D, Patel M, Hunter AM, Barber DE, Kobet GL (2009) Preliminary synchronized phasor data analysis of disturbance events in the US eastern interconnection. In: 2009 IEEE/PES power systems conference and exposition, pp 1–8. http://dx.doi.org/10.1109/PSCE.2009.4840097
2. Amin S, Cárdenas AA, Sastry SS (2009) Safe and secure networked control systems under denial-of-service attacks. In: Majumdar R, Tabuada P (eds) Hybrid systems: computation and control. Springer, Berlin, pp 31–45
3. Cárdenas AA, Amin S, Sastry S (2008) Research challenges for the security of control systems. In: Proceedings of the 3rd conference on hot topics in security, HOTSEC'08, USENIX Association, pp 1–6
4. Cárdenas AA, Amin S, Lin Z-S, Huang Y-L, Huang C-Y, Sastry S (2011) Attacks against process control systems: risk assessment, detection, and response. In: Proceedings of the 6th ACM symposium on information, computer and communications security, ASIACCS '11, Association for computing machinery, pp 355–366. http://dx.doi.org/10.1145/1966913.1966959
5. Cardenas AA, Amin S, Sastry S (2008) Secure control: towards survivable cyber-physical systems. In: 2008 The 28th international conference on distributed computing systems workshops, pp 495–500. http://dx.doi.org/10.1109/ICDCS.Workshops.2008.40
6. Chamorro HR, Sanchez AC, Pantoja A, Zelinka I, Gonzalez-Longatt F, Sood VK (2019) A network control system for hydro plants to counteract the non-synchronous generation integration. Int J Electrical Power Energy Syst 105:404–419. http://dx.doi.org/https://doi.org/10.1016/j.ijepes.2018.08.020. http://www.sciencedirect.com/science/article/pii/S0142061517323566
7. Mirez JL, Chamorro HR, Ordonez CA, Moreno R (2014) Energy management of distributed resources in microgrids. In: 2014 IEEE 5th Colombian workshop on circuits and systems (CWCAS), 2014, pp 1–5
8. Moreno R, Chamorro HR, Izadkhast SM (2013) A framework for the energy aggregator model. In: Workshop on power electronics and power quality applications (PEPQA) 2013, pp 1–5
9. Liu Y, Ning P, Reiter MK (2009) False data injection attacks against state estimation in electric power grids. In: Proceedings of the 16th ACM conference on computer and communications security, CCS '09, Association for computing machinery, pp 21–32. http://dx.doi.org/10.1145/1653662.1653666
10. Dán G, Sandberg H (2010) Stealth attacks and protection schemes for state estimators in power systems. In: 2010 first IEEE international conference on smart grid communications, pp 214–219. http://dx.doi.org/10.1109/SMARTGRID.2010.5622046
11. Kosut O, Jia L, Thomas RJ, Tong L (2010) Malicious data attacks on smart grid state estimation: attack strategies and countermeasures. In: 2010 first IEEE international conference on smart grid communications, pp 220–225. http://dx.doi.org/10.1109/SMARTGRID.2010.5622045
12. Sairam T, Swarup KS (2017) Analysis of network topology processor algorithms in substation level networks. In: 2017 7th international conference on power systems (ICPS), 2017, pp 601–606
13. Lin J, Yu W, Yang X (2016) Towards multistep electricity prices in smart grid electricity markets. IEEE Trans Parallel Distrib Syst 27(1):286–302
14. Xie L, Mo Y, Sinopoli B (2010) False data injection attacks in electricity markets. In: 2010 first IEEE international conference on smart grid communications, pp 226–231. http://dx.doi.org/10.1109/SMARTGRID.2010.5622048
15. Phillips LR, Tejani B, Margulies J, Hills JL, Richardson BT, Baca MJ, Weiland L (2005) Analysis of operations and cyber security policies for a system of cooperating flexible alternating current transmission system (FACTS) devices. http://dx.doi.org/10.2172/882347
16. Sridhar S, Manimaran G (2011) Data integrity attack and its impacts on voltage control loop in power grid. In: 2011 IEEE power and energy society general meeting, pp 1–6. http://dx.doi.org/10.1109/PES.2011.6039809

17. Esfahani PM, Vrakopoulou M, Margellos K, Lygeros J, Andersson G (2010) A robust policy for automatic generation control cyber attack in two area power network. In: 49th IEEE conference on decision and control (CDC), pp 5973–5978. http://dx.doi.org/10.1109/CDC.2010.5717285
18. Esfahani PM, Vrakopoulou M, Margellos K, Lygeros J, Andersson G (2010) Cyber attack in a two-area power system: impact identification using reachability. In: Proceedings of the 2010 American control conference, pp 962–967. http://dx.doi.org/10.1109/ACC.2010.5530460
19. Pasqualetti F, Dörfler F, Bullo F (2013) Attack detection and identification in cyber-physical systems. IEEE Trans Autom Control 58(11):2715–2729

Chapter 7
Conclusions

This book has focused on key issues related to the structural properties of the power system network topology and the advantageous usage of this information to perform security studies. In this chapter, we summarize the work presented in this book and discuss some problems that are extensions of the reported research results and some possible directions for future research.

We have developed in Chap. 2 a comprehensive topological characterization of the power system networks that explicitly considers all aspects of its network topology. The mathematical characterization uses extensively graph theory concepts to capture all properties of the power system networks. We use the connectivity matrix and the degree matrix to construct the Laplacian matrix that represents the interconnection among substations through the transmission system. The mathematical structure of the Laplacian matrix and its properties are shared with a modified version of the admittance matrix. The topological characterization has served as the basis of the formulation of various issues treated in the book.

Chapter 3 has presented the mathematical formulation based on the topological characterization to identify multiple subnetworks in the power systems under some restrictions. The subnetworks identified are connected with a minimum number of transmission lines. If the modified version of the admittance matrix is used then the transfer admittance among subnetwork is minimal. We develop the formulation of an unconstrained optimization problem and the solution of this problem allows us to identify directly k subnetworks. The development has shown that the identification problem is transformed into an eigenvalue and eigenvector problem involving the graph Laplacian.

Chapter 4 has presented a comprehensive development to quantify the network robustness of power system networks. We provide a concept of network robustness based on the topological firmness of the interconnection among substations through of the transmission system. We discussed about the relative importance of the buses inside the networks and how some nodes play an important role in the connectivity of the whole network. The index deduced to quantify the network robustness is useful to take decisions about the planning of the transmission system; for instance, to take decisions about where should be reinforced the network.

We have developed a security strategy for power systems in Chap. 5 based on the identification of multiple subnetworks. We call this strategy as "Hierarchical Islanding (HI)". The strategy was developed as a software tool based mainly on two modules: Topological Module (TM) and Spectral Analysis Module (SAM). The strategy offers the possibility to power system operators to eject an islanding-action to mitigate grid-impactive events. The knowledge of the sets of transmission lines that connects subnetworks allows the reduction of widespread events based on the request by the user to create two islands.

We have discussed in the Chap. 6 about a recent topic in power system which is object of a lot of research effort. The Cyber Physical Security of the power systems is a challenging topic given the highly dependence of the modern society to the continuous service of electricity. The integration of Information Technology (IT) systems (computations and communication-the cyber world) with sensor and actuation data (the physical world) has introduced new and fundamentally different approaches to security research. We provide a deeply discussion about the specialized literature proposing different approaches to study how attacks and resilient mechanisms can affect and defend power grid operations.

7.1   Future Research Directions

While the development of the models and the proposed approaches provide tools that therefore did not exist, there are numerous modifications and/or enhancements that can be introduced to overcome certain identifiable shortcomings and to extend the capabilities of the analysis and the tools.

The Graph Theory and the most recent topic Complex Networks offer numerous analyses to study networks from different perspectives. Therefore, and particularly, the study of the power systems networks from this perspective is still in the beginning. In this book, we provide the deduction of important structural characteristics but there are more properties ones that can be researched or extended. The challenge in this point is on a precise approach to study power system networks in terms of useful outcomes. For instance, the topological analysis of the power grid can provide valuable information about the expansion of the transmission system during planning studies. The current planning studies take into account common reliability measures but the there are other measures such as network properties that can provide useful information. We think that the inclusion of network analysis into planning studies is a challenging field of research.

On the other hand, despite of the recent advance in complex networks to understand the performance of networks, our understanding of cascading failures is rather limited. Topological robustness is a structural feature of networks. Cascading failures, however, are a dynamic property of complex systems.

In terms of networks, there is yet another perspective to study power system networks considering the nature of the graph. In this book, our interest is on the topological structure; however, the characterization of the graph can be modified in three manners. First, the usage of active and reactive power serves to characterize the graph that represents the power systems. Second, the usage of market information to characterize the graph, for instance, signal prices such as Locational Marginal Price (LMP) can be used to assign values to each transaction in order to deduce useful information about the electricity markets. Third, the relationship among dynamical variables can be deduced using appropriately graph models.

Work on fault tolerance and reliability of control system is not enough, because these mechanisms generally assume independent and uncorrelated failures; however, cyber-attacks will exploit vulnerabilities in a coordinated and correlated fashion.

In the promising area of cyber physical security, the work on fault tolerance and reliability of control system is not enough, because these mechanisms generally assume independent and uncorrelated failures; however, cyber-attacks will exploit vulnerabilities in a coordinated and correlated fashion. Therefore instead of relying solely on fault-detection algorithms to protect control algorithms in the power grid, we need to develop new attacks-detection algorithms focusing on identifying malicious data in sensor actuation devices in the power grid.

The suggested topics mentioned before provide some fruitful directions for future research in extending the current analysis and tools so as to develop additional capabilities for more comprehensive assessments.

Appendix A

A.1 Proposition 1

The expression:

$$|\mathcal{L}_{12}| = \frac{1}{2} \sum_{i,j \in N} (r_i - r_j) \tag{A.1}$$

Could be written in term of the matrix $\mathbf{L}$

$$\frac{1}{2} \sum_{i,j \in N} (r_i - r_j)^2 = \mathbf{r}^T \mathbf{L} \mathbf{r} \tag{A.2}$$

Development:

$$|\mathcal{L}_{12}| = \frac{1}{2} \sum_i \sum_j (r_i - r_j)^2 \tag{A.3}$$

$$|\mathcal{L}_{12}| = \frac{1}{2} \sum_i \sum_j \left(r_i^2 - 2 r_i r_j + r_j^2 \right) \tag{A.4}$$

$$|\mathcal{L}_{12}| = \frac{1}{2} \left(\sum_i r_i^2 - 2 \sum_i \sum_j r_i r_j + \sum_j r_j^2 \right) \tag{A.5}$$

$$|\mathcal{L}_{12}| = \frac{1}{2} \left(\sum_i r_i^2 - 2 \sum_i \sum_{j \neq i} r_i r_j \right) \tag{A.6}$$

R. Moreno Chuquen et al., *Graph Theory Applications to Deregulated Power Systems*,
SpringerBriefs in Electrical and Computer Engineering,
https://doi.org/10.1007/978-3-030-57589-2

Appendix B

B.1 Proposition 2

The analytical development for the quadratic optimization problem to identify the set of transmission lines connecting two subnetworks:

$$|\mathcal{L}_{12}| = \frac{1}{2} \left(\sum_i r_i^2 - 2 \sum_i \sum_{j \neq i} r_i r_j \right) \tag{B.1}$$

s.t.

Solve (B.1)–(B.3) as an unconstrained optimization for introducing the Lagrange multipliers λ_1 for $h(\mathbf{r}) = 0$ and λ_2 for $g(\mathbf{r}) = 0$ The necessary conditions of optimality are:

$$h(\mathbf{r}) = 0 \tag{B.2}$$

$$g(\mathbf{r}) = 0 \tag{B.3}$$

The differentiation with respect to $\mathbf{r}$ yields:

$$\mathbf{Lr} - \lambda_1 \mathbf{1} - \lambda_2 \mathbf{r} = 0 \tag{B.4}$$

or,

$$(\mathbf{L} - \lambda_2 \mathbf{I}) \mathbf{r} = \lambda_1 \mathbf{r} \tag{B.5}$$

The multiplication by $\mathbf{1}^T$ on both sides of (B.5) obtains

$$\mathbf{1}^T \mathbf{Lr} - \lambda_2 \mathbf{1}^T = \lambda_1 \mathbf{1}^T \mathbf{1} = N \lambda_1 \tag{B.6}$$

© The Author(s), under exclusive license to Springer Nature Switzerland AG 2021
R. Moreno Chuquen et al., *Graph Theory Applications to Deregulated Power Systems*,
SpringerBriefs in Electrical and Computer Engineering,
https://doi.org/10.1007/978-3-030-57589-2

Since, $\mathbf{1}^T\mathbf{L} = 0$ and $\mathbf{1}^T\mathbf{r} = 0$, so $\lambda_1 = 0$, thus

$$\mathbf{1}^T\mathbf{L}\mathbf{r} - \lambda_2\mathbf{1}^T = \lambda_1\mathbf{1}^T\mathbf{1} = N\lambda_1 \qquad \text{(B.7)}$$

Printed in the United States
By Bookmasters